Getting Started
with the
TI-83 Plus

Important information

Texas Instruments makes no warranty, either express or implied, including but not limited to any implied warranties of merchantability and fitness for a particular purpose, regarding any programs or book materials and makes such materials available solely on an "as-is" basis.

In no event shall Texas Instruments be liable to anyone for special, collateral, incidental, or consequential damages in connection with or arising out of the purchase or use of these materials, and the sole and exclusive liability of Texas Instruments, regardless of the form of action, shall not exceed the purchase price of this calculator. Moreover, Texas Instruments shall not be liable for any claim of any kind whatsoever against the use of these materials by any other party.

US FCC Information Concerning Radio Frequency Interference

This equipment has been tested and found to comply with the limits for a Class B digital device, pursuant to Part 15 of the FCC rules. These limits are designed to provide reasonable protection against harmful interference in a residential installation. This equipment generates, uses, and can radiate radio frequency energy and, if not installed and used in accordance with the instructions, may cause harmful interference with radio communications. However, there is no guarantee that interference will not occur in a particular installation.

If this equipment does cause harmful interference to radio or television reception, which can be determined by turning the equipment off and on, you can try to correct the interference by one or more of the following measures:

- Reorient or relocate the receiving antenna.
- Increase the separation between the equipment and receiver.
- Connect the equipment into an outlet on a circuit different from that to which the receiver is connected.
- Consult the dealer or an experienced radio/television technician for help.

Caution: Any changes or modifications to this equipment not expressly approved by Texas Instruments may void your authority to operate the equipment.

© 2000–2003, 2007 Texas Instruments Incorporated

Windows, Macintosh are the property of their respective owners.

Contents

About this book 1
TI-83 Plus keys 2
Turning the TI-83 Plus on and off 3
Home screen 4
[2nd] and [ALPHA] keys 6
[CLEAR] and [2nd] [QUIT] 7
Entering an expression 8
TI-83 Plus menus 9
Editing and deleting 12
Using [-] and [(-)] 14
Using parentheses 16
Storing a value 18
Graphing a function 20
Changing mode settings 22
Setting the graphing window 25
Using [ZOOM] 27
Building a table 28
Using the CATALOG 30
Performing simple calculations 31
Using the equation solver 34
Entering data into lists 36
Plotting data 38
Calculating a linear regression 42

iii

Calculating statistical variables 43

Using the MATRIX Editor 44

Grouping 48

Ungrouping 50

Error messages 51

Resetting defaults 52

Installing applications 53

Running applications 54

Quick reference 55

Texas Instruments (TI) Support and Service 56

Warranty information 57

Battery precautions 59

About this book

This *Getting Started Guide* was designed for:

- students who are using a graphing calculator for the first time.
- students who need a quick review of procedures for common operations on the TI-83 Plus.

This book gives a quick overview of each topic, along with keystroke instructions for easy examples. All examples assume that the TI-83 Plus is using default settings.

For complete information on any topic, see the electronic Guidebook on the CD that came with your calculator. The electronic Guidebook is a complete reference manual for the TI-83 Plus. If the CD is not available, you can download a copy of the electronic guidebook from the Texas Instruments web page at

education.ti.com/guides

Look for the 📖 symbol at the top of the page in this guide. These notes direct you to the chapter in the electronic Guidebook that provides complete details about the topic.

The TI-83 Plus also has some calculator software applications (APPS) preinstalled.

For information about these APPS, see the electronic documentation files on the Texas Instrument web page at

education.ti.com/guides

TI-83 Plus keys

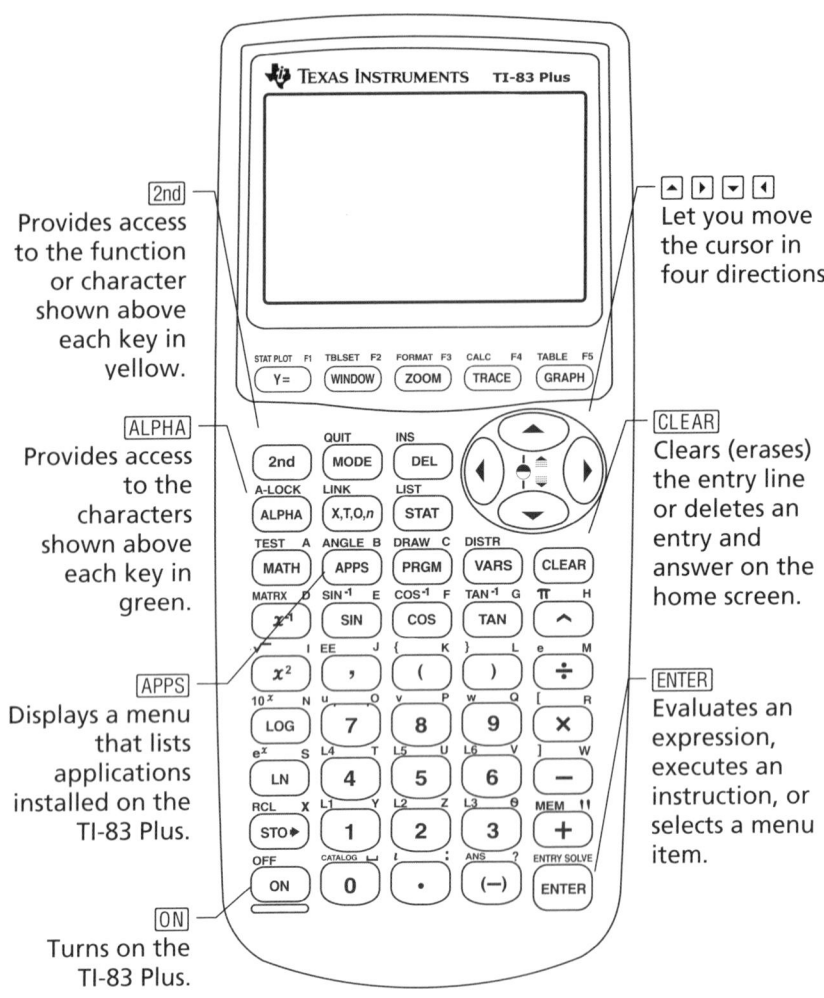

[2nd]
Provides access to the function or character shown above each key in yellow.

[ALPHA]
Provides access to the characters shown above each key in green.

[APPS]
Displays a menu that lists applications installed on the TI-83 Plus.

[ON]
Turns on the TI-83 Plus.

[▲][▶][▼][◀]
Let you move the cursor in four directions.

[CLEAR]
Clears (erases) the entry line or deletes an entry and answer on the home screen.

[ENTER]
Evaluates an expression, executes an instruction, or selects a menu item.

2

Turning the TI-83 Plus on and off

To turn on the TI-83 Plus, press [ON]. The [ON] key is located at the lower left corner of the TI-83 Plus.

For more details, see Guidebook Chapter 1.

To turn off the TI-83 Plus, press the [2nd] key followed by the [ON] key. OFF is the *second* function of [ON].

When you turn off the TI-83 Plus, all settings and memory contents are retained. The next time you turn on the TI-83 Plus, the home screen displays as it was when you last used it.

Automatic Power Down™

To prolong the life of the batteries, Automatic Power Down™ (APD™) turns off the TI-83 Plus automatically after about five minutes without any activity. The next time you turn on the calculator, it is exactly as you left it.

Home screen

When you turn on your TI-83 Plus the first time, you should see this screen:

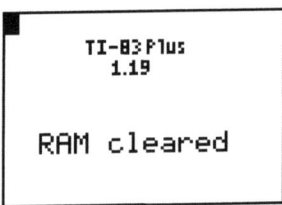

To clear this text from your screen, press [CLEAR] twice. You should now see the home screen, a blank screen with a flashing cursor. The home screen is where you enter problems and see results.

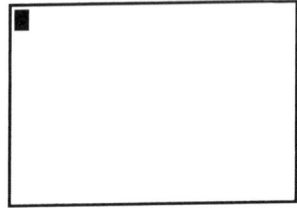

If you pressed [CLEAR] above and you still do not see a blank home screen, press the [2nd] key followed by the [MODE] key (to select QUIT).

Home screen (continued)

Example: Add 2 + 3 on the home screen.

Press	Result
2 [+] 3	`2+3`■
[ENTER]	`2+3` `          5` ← Entry line / Answer line

Note: Results are displayed on the next line (the answer line), not on the entry line.

Example: Multiply 5 x 4.

Press	Result

5 [×] 4
[ENTER]
	`5*4`
	`         20`
	■

[2nd] and [ALPHA] keys

Most keys on the TI-83 Plus can perform two or more functions. To use a function printed on a key, press the key. To use a function printed above a key in yellow or green, you must first press the [2nd] key or the [ALPHA] key.

[2nd] key

Second functions are printed above the keys in yellow (the same color as the [2nd] key). Some secondary functions enter a function or a symbol on the home screen (sin^{-1} or $\sqrt{\ }$, for example). Others display menus or editors.

To view the ANGLE menu, for example, look for ANGLE (printed in yellow) above the blue [APPS] key near the top of the TI-83 Plus keyboard. Press the [2nd] key (and then release it) and then press [APPS]. In this book and in the TI-83 Plus Guidebook, this key combination is indicated by [2nd] [ANGLE], not [2nd] [APPS].

Note: The flashing cursor changes to ◘ when you press the [2nd] key.

[ALPHA] key

The [ALPHA] key lets you enter the alphabetic characters and some special symbols. To enter T, for example, press [ALPHA] (and then release it) and then press [4]. In this book and in the TI-83 Plus Guidebook, this key combination is indicated by [ALPHA] [T].

If you have several alphabetic characters to enter, press [2nd] [A-LOCK] to avoid having to press the [ALPHA] key multiple times. This locks the alpha key in the *On* position until you press [ALPHA] a second time to unlock it.

Note: The flashing cursor changes to ◘ when you press the [ALPHA] key.

[CLEAR] and [2nd] [QUIT]

[CLEAR] key

The [CLEAR] key erases the home screen. This key is located just below the four blue arrow keys at the upper right corner of the TI-83 Plus keyboard. If you press [CLEAR] during an entry, it clears the entry line. If you press [CLEAR] when the cursor is on a blank line, it clears everything on the home screen.

Although it does not affect the calculation, it is frequently helpful to clear the previous work from the home screen before you begin a new problem. As you work through this guide, we recommend that you press [CLEAR] each time you begin a new *Example*. This removes the previous example from the home screen and ensures that the screen you see matches the one shown in the example.

[2nd] [QUIT]

If you accidentally press a menu key, pressing [CLEAR] will sometimes return you to the home screen, but in most cases you must press [2nd] [QUIT] to leave the menu and return to the home screen.

Entering an expression

An expression consists of numbers, variables, operators, functions, and their arguments that evaluate to a single answer. 2X + 2 is an expression.

Type the expression, and then press [ENTER] to evaluate it. To enter a function or instruction on the entry line, you can:

- Press its key, if available. For example, press [LOG].

  — or —

- Select it from the CATALOG, if the function appears on the CATALOG. For example, press [2nd] [CATALOG], press [▼] to move down to **log(**, and press [ENTER] to select **log(**.

  — or —

- Select it from a menu, if available. For example, to find the **round** function, press [MATH], press [▶] to select **NUM**, then select **2:round(**.

Example: Enter and evaluate the expression π × 2.

Press	Result
[2nd] [π] [×] 2	π*2
[ENTER]	π*2 6.283185307

TI-83 Plus menus

Many functions and instructions are entered on the home screen by selecting from a menu.

For more details, see Guidebook Chapter 1.

To select an item from the displayed menu:

- Press the number or letter shown at the left of that item.
  — or —
- Use the cursor arrow keys, ▼ or ▲, to highlight the item, and then press [ENTER].

Some menus close automatically when you make a selection, but if the menu remains open, press [2nd] [QUIT] to exit. Do not press [CLEAR] to exit, since this will sometimes delete your selection.

Example: Enter $\sqrt[3]{27}$ on the home screen entry line.

Press	Result
[MATH]	**MATH** NUM CPX PRB 1:▶Frac 2:▶Dec 3:³ 4:³√(5:ˣ√ 6:fMin(7↓fMax(Menus containing an arrow next to the final item continue on a second page.
4 — or — ▼ ▼ ▼ [ENTER]	³√(
2 7) [ENTER]	³√(27) 3

9

TI-83 Plus menus (continued)

Example: Change the FORMAT menu setting to display grid points on the graph.

Press	Result
[2nd] [FORMAT]	RectGC PolarGC CoordOn CoordOff GridOff GridOn AxesOn AxesOff LabelOff LabelOn ExprOn ExprOff
[▼] [▼] [▶] [ENTER]	RectGC PolarGC CoordOn CoordOff GridOff **GridOn** AxesOn AxesOff LabelOff LabelOn ExprOn ExprOff
[GRAPH]	

Example: Turn off the display of grid points.

Press	Result
[2nd] [FORMAT] [▼] [▼] [ENTER]	RectGC PolarGC CoordOn CoordOff GridOff GridOn AxesOn AxesOff LabelOff LabelOn ExprOn ExprOff

Note: Press [2nd] [QUIT] *or* [CLEAR] *to close the FORMAT menu and return to the home screen.*

10

TI-83 Plus menus (continued)

Summary of menus on the TI-83 Plus

Press	To display
[APPS]	APPLICATIONS menu — to see a list of TI-83 Plus calculator software applications (APPS).
[2nd] [LINK]	LINK menu — to communicate with another calculator.
[2nd] [MEM]	MEMORY menu — to check available memory and manage existing memory.
[MATH]	MATH menu — to select a math operation.
[VARS]	VARS menu — to select variable names to paste to the home screen.
[2nd] [STAT PLOT]	STAT PLOTS menu — to define statistical plots.
[2nd] [CATALOG]	CATALOG menu — to select from a complete, alphabetic list of all TI-83 Plus built-in functions and instructions.
[2nd] [FORMAT]	FORMAT menu — to define a graph's appearance.
[2nd] [MATRIX]	MATRIX menu — to define, view, and edit matrices.
[2nd] [DRAW]	DRAW menu — to select tools for drawing on graphs.
[2nd] [DISTR]	DISTRIBUTIONS menu — to select distribution functions to paste to the home screen or editor screens.
[2nd] [TEST]	TEST menu — to select relational operators (=, ≠, ≤, ≥, etc.) and Boolean operators (and, or, xor, not) to paste to the home screen.

Editing and deleting

You can change any expression or entry using the backspace ◀ key, the delete [DEL] key, or the insert [2nd] [INS] key. You can make a change before or after you press [ENTER].

Example: Enter the expression $5^2 + 1$, and then change the expression to $5^2 + 5$.

Press	Result
5 [x^2] [+] 1	5^2+1■
◀ 5	5^2+5

Example: Enter the expression $5^2 + 1$, and then change the expression to $5^2 - 5$.

Press	Result
5 [x^2] [+] 1	5^2+1■

Editing and deleting (continued)

Press	Result
[◄] [◄] [DEL] [DEL]	5^2
[-] 5 [ENTER]	5^2-5 20

Example: Change the example above to $5^2 + 2 - 5$ using [2nd] [ENTRY] to recall the expression and [2nd] [INS] to insert + 2 into the expression.

Press	Result
[2nd] [ENTRY]	5^2-5▮
[◄] [◄] [2nd] [INS] [+] 2 [ENTER]	5^2+2-5 22

Using ⊟ and ⍺

Many calculators (including the TI-83 Plus) make a distinction between the symbols for subtraction and negation.

Use ⊟ to enter subtraction operations. Use ⍺ to enter a negative number in an operation, in an expression, or on a setup screen.

Example: Subtract 10 from 25.

Press	Result
2 5 ⊟ 1 0 [ENTER]	25-10 15

Example: Add 10 to -25.

Press	Result
⍺ 2 5 ⊞ 1 0 [ENTER]	-25+10 -15

Using ⊟ and ⍐ (continued)

Example: Subtract ⁻10 from 25.

Press	Result
2 5 ⊟ ⍐ 1 0 [ENTER]	25--10 35

Note: Notice that the TI-83 Plus displays a slightly different symbol for negation and subtraction to make it easier for you to distinguish between the two. The negative symbol is raised and slightly shorter.

Using parentheses

Since all calculations inside parentheses are completed first, it is sometimes important to place a portion of an expression inside parentheses.

For more details, see Guidebook Chapter 3.

Example: Multiply 4*1+2; then multiply 4*(1+2).

Press	Result
4 [×] 1 [+] 2 [ENTER]	4*1+2 6
4 [×] [(] 1 [+] 2 [)] [ENTER]	4*1+2 6 4*(1+2) 12

Note: The closing parenthesis [)] is optional. The operation will be completed if you omit it. The exception to this rule occurs when there is another operation following the parenthetical operation. In this case, you must include the closing parenthesis.

Using parentheses (continued)

Example: Divide 1/2 by 2/3.

Press	Result

`(` 1 `÷` 2 `)` `÷`
 `(` 2 `÷` 3 `)`
 `ENTER`
	(1/2)/(2/3)
	 .75

Example: Calculate $16 \wedge \frac{1}{2}$.

Press	Result
1 6 `^` `(` 1 `÷` 2 `)` `ENTER`	16^(1/2) 4

Example: Calculate $(-3)^2$.

Press	Result
`(` `(-)` 3 `)` `x²` `ENTER`	(-3)² 9

Note: *Try each of these examples without the parentheses and see what happens!*

Storing a value

Values are stored to and recalled from memory using variable names.

Example: Store 25 to variable A and multiply A by 2.

Press	Result
2 5 [STO▶] [ALPHA] [A]	25→A
[ENTER]	25→A 25
2 [×] [ALPHA] [A] [ENTER]	25→A 25 2∗A 50
— or — [ALPHA] [A] [×] 2 [ENTER]	25→A 25 2∗A 50 A∗2 50

Storing a value (continued)

Example: Find the value of $2X^3 - 5X^2 - 7X + 10$ when $X = {}^-0.5$.

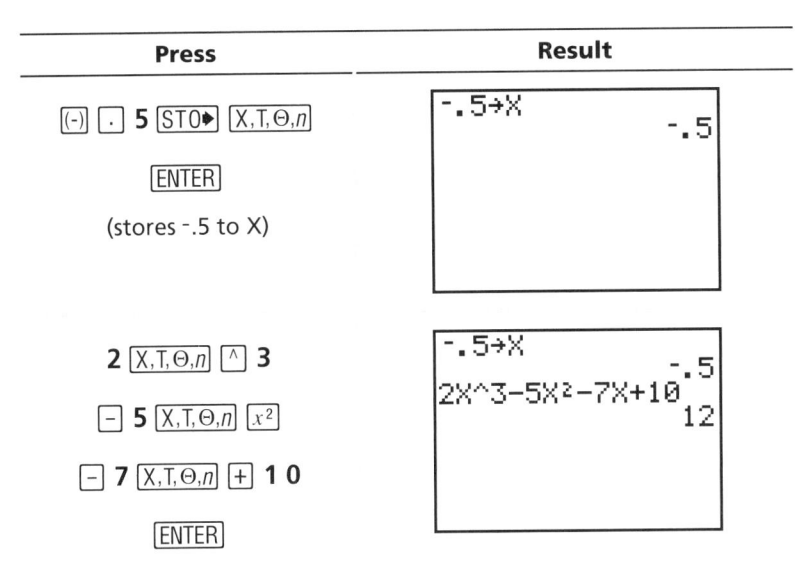

Press	Result
[(-)] [.] 5 [STO▶] [X,T,Θ,n] [ENTER] (stores -.5 to X)	`-.5→X` `           -.5`

2 [X,T,Θ,n] [^] 3
[-] 5 [X,T,Θ,n] [x^2]
[-] 7 [X,T,Θ,n] [+] 1 0
[ENTER]
	`-.5→X`
	`           -.5`
	`2X^3-5X²-7X+10`
	`            12`

You can remove a value stored to a variable using the DELVAR function or by storing 0 to the variable.

Example: Delete the value (-.5) stored to X above by storing 0.

Press	Result
0 [STO▶] [X,T,Θ,n] [ENTER]	`0→X` `             0`

[X,T,Θ,n]
[ENTER]
	`0→X`
	`             0`
	`X`
	`             0`

19

Graphing a function

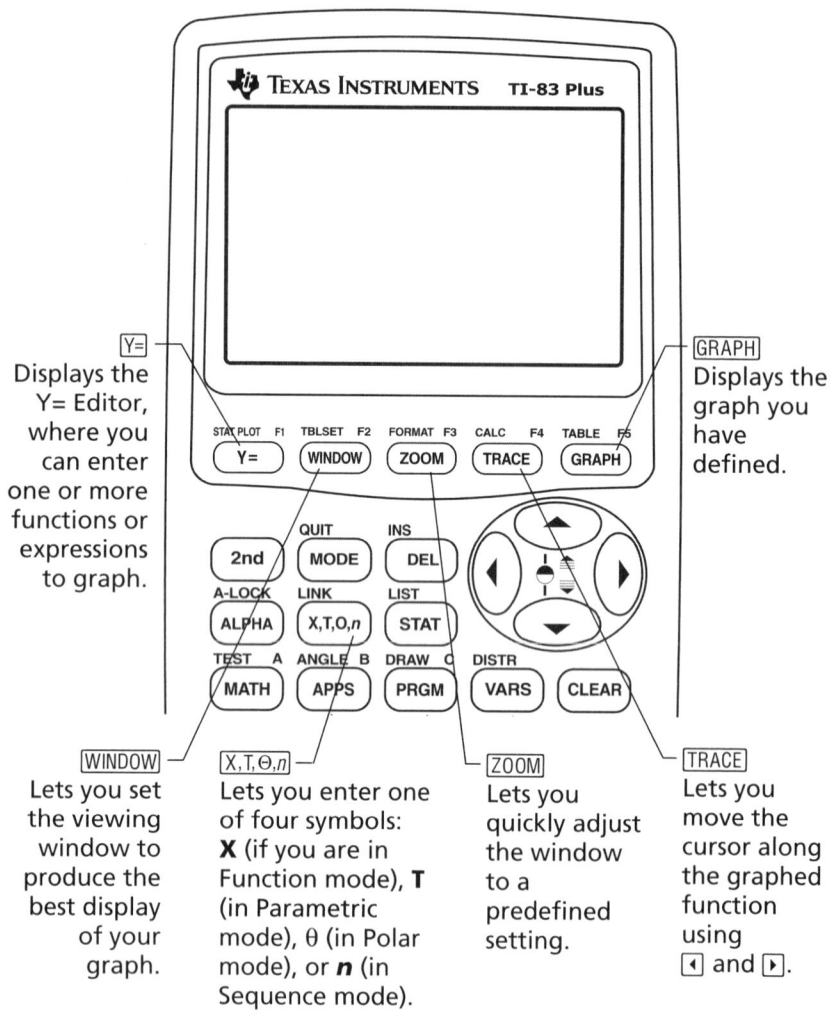

[Y=]
Displays the Y= Editor, where you can enter one or more functions or expressions to graph.

[GRAPH]
Displays the graph you have defined.

[WINDOW]
Lets you set the viewing window to produce the best display of your graph.

[X,T,Θ,n]
Lets you enter one of four symbols: **X** (if you are in Function mode), **T** (in Parametric mode), θ (in Polar mode), or **n** (in Sequence mode).

[ZOOM]
Lets you quickly adjust the window to a predefined setting.

[TRACE]
Lets you move the cursor along the graphed function using [◄] and [►].

20

Graphing a function (continued)

To graph a function, you must:

1. Display the Y= Editor.
2. Enter the function.
3. Display the graph.

📖 For more details, see Guidebook Chapter 3.

Note: *If you previously changed graph type in the mode settings, you must change the type back to Func (the default setting) before you graph.*

Example: Graph the function Y = X² + 1.

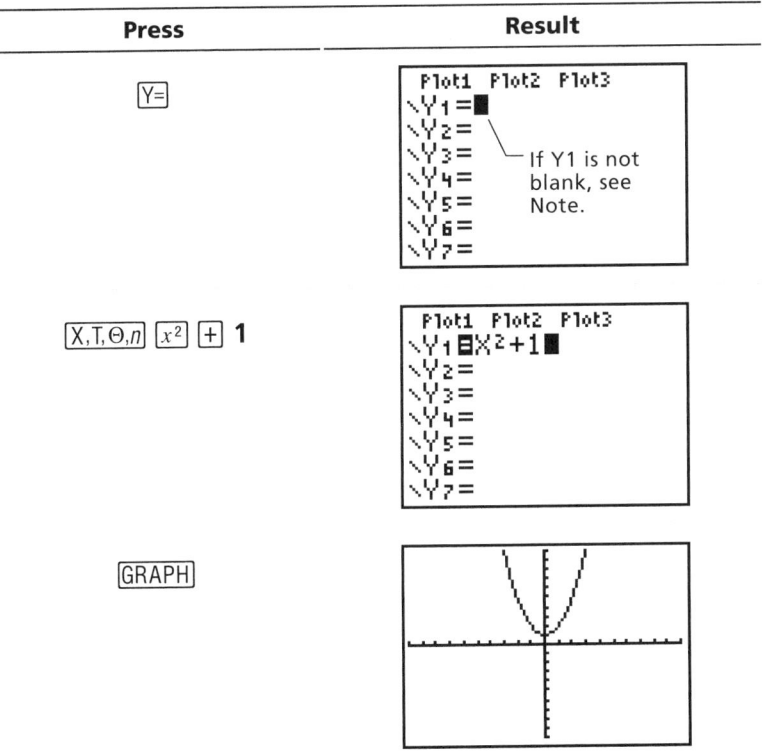

Note: *If Y1 is not empty, press* [CLEAR]. *If there are additional entries in the Y= Editor, press* [▼] [CLEAR] *until all are clear.*

21

Changing mode settings

The mode settings determine how entries are interpreted and how answers are displayed on the TI-83 Plus.

📖 For more details, see Guidebook Chapter 1.

Example: Change the mode setting for decimals from *Float* to 3 decimal places.

Press	Result
[MODE]	**Normal** Sci Eng **Float** 0123456789 **Radian** Degree **Func** Par Pol Seq **Connected** Dot **Sequential** Simul **Real** a+bi re^θi **Full** Horiz G-T
▼ ▶ ▶ ▶ ▶ [ENTER]	**Normal** Sci Eng Float 012**3**456789 **Radian** Degree **Func** Par Pol Seq **Connected** Dot **Sequential** Simul **Real** a+bi re^θi **Full** Horiz G-T
[2nd] [QUIT] 1 [.] 2 3 4 5 6 [ENTER]	1.23456 1.235

Note: *You must press* [ENTER] *to change a mode setting. If you highlight the setting and then exit the mode menu without pressing* [ENTER]*, the setting will not be changed.*

Changing mode settings (continued)

The mode menu includes the following settings:

Setting	Choices
Numeric notation	*Normal:* for example, 12345.67 *Sci* (scientific): for example, 1.234567E4 *Eng* (engineering): for example, 12.34567E3
Decimal	*Float:* lets the number of decimal places change based on the result (up to 10 digits) *0 - 9:* sets the number of decimal places to a value (0 - 9) that you specify
Angle measure	*Radian:* interprets angle values as radians *Degree:* interprets angle values as degrees
Type of graph	*Func* (functional): plots functions, where Y is a function of X *Par* (parametric): plots relations, where X and Y are functions of T *Pol* (polar): plots functions, where r is a function of $[n]\theta$ *Seq* (sequence): plots sequences
Plot type	*Connected:* draws a line connecting each point calculated for the selected functions *Dot:* plots only the calculated points of the selected functions
Sequential or simultaneous graphing	*Sequential:* draws graphs one at a time *Simul* (simultaneous): draws several graphs at the same time
Real or complex mode	*Real:* displays real numbers, such as 1, 1/2, $\sqrt{3}$ *a+bi* (rectangular complex): displays as 3+2i *re^θi* (polar complex): displays as re^θi
Screen display	*Full:* displays full screen *Horiz:* displays a horizontal split screen *G-T:* displays a vertical split screen (graph & table)

Changing mode settings (continued)

The importance of mode settings

Example: Multiply 2/3 × 2.

Press	Result
[MODE] [▼] [▶] [ENTER]	**Normal** Sci Eng **Float** 0123456789 **Radian** Degree **Func** Par Pol Seq **Connected** Dot **Sequential** Simul **Real** a+bi re^θi **Full** Horiz G-T
2 [÷] 3 [×] 2 [ENTER]	2/3*2 　　　　1

Your first reaction to this example is that the calculator has produced a wrong answer. But you have set it to round to 0 decimal places (the nearest whole number), so for this setting the answer is correct. If you set rounding (decimals displayed) to 0 and then forget to reset it for later calculations, you may be surprised by some of your answers! With mode set to the default setting of *Float*, the result will be:

Press	Result
2 [÷] 3 [×] 2 [ENTER]	2/3*2 　　　1.333333333

Setting the graphing window

To obtain the best view of the graph, you may need to change the boundaries of the window.

📖 For more details, see Guidebook Chapter 3.

To display the WINDOW Editor, press [WINDOW].

Window variables
(shown in WINDOW Editor)

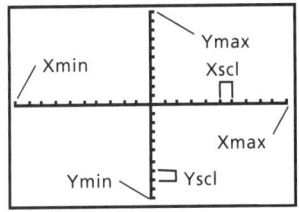

Corresponding viewing window
(shown on Graph window)

The *Xmin*, *Xmax*, *Ymin*, and *Ymax* variables represent the boundaries of the viewing window.

Xmin: the minimum value of X to be displayed.
Xmax: the maximum value of X to be displayed.
Ymin: the minimum value of Y to be displayed.
Ymax: the maximum value of Y to be displayed.
Xscl (X scale): the distance between the tick marks on the X axis.
Yscl (Y scale): the distance between the tick marks on the Y axis.
Xres: pixel resolution—not usually changed except by advanced users.

To change the values:

1. Move the cursor to highlight the value you want to change.
2. Do one of the following:
   - Type a value or an expression. The old value is erased when you begin typing.

     — or —

   - Press [CLEAR] to clear the old value; then type the new one.
3. Press [ENTER], [▼], or [▲].

Setting the graphing window (continued)

Note: *Values are stored as you type them; you do not need to press [ENTER]. Pressing [ENTER] simply moves the cursor to the next window variable.*

4. After you have made all changes, press [2nd] [QUIT] to close the WINDOW Editor (or [GRAPH] to display the graph).

Example: Change the window settings to display a maximum X value of 25, a minimum X value of -25, a maximum Y value of 50, and a minimum Y value of -50.

Press	Result
[WINDOW]	WINDOW Xmin=-10 Xmax=10 Xscl=1 Ymin=-10 Ymax=10 Yscl=1 Xres=1
[(-)] 2 5 [▼] 2 5 [▼] [▼] [(-)] 5 0 [▼] 5 0	WINDOW Xmin=-25 Xmax=25 Xscl=1 Ymin=-50 Ymax=50 Yscl=1 Xres=1
[2nd] [QUIT]	■

Using [ZOOM]

The TI-83 Plus has ten predefined window settings that let you quickly adjust the graph window to a predetermined level of magnification. To display this menu, press [ZOOM].

📖 For more details, see Guidebook Chapter 3.

Selection	Result
1: ZBox	Lets you draw a box (using the cursor pad) to define the viewing window.
2: Zoom In	After you position the cursor and press [ENTER], magnifies the graph around the cursor.
3: Zoom Out	After you position the cursor and press [ENTER], displays more of the graph.
4: ZDecimal	Sets the change in X and Y to increments of .1 when you use [TRACE].
5: ZSquare	Adjusts the viewing window so that X and Y dimensions are equal.
6: ZStandard	Sets the standard (default) window variables.
7: ZTrig	Sets the built-in trigonometry window variables.
8: ZInteger	After you position the cursor and press [ENTER], sets the change in X and Y to whole number increments.
9: ZoomStat	Sets the values for currently defined statistical lists.
0: ZoomFit	Fits **Ymin** and **Ymax** between **Xmin** and **Xmax**.

Building a table

Tables are useful tools for comparing values for a function at multiple points.

For more details, see Guidebook Chapter 7.

Example: Build a table to evaluate the function Y = X³ - 2X at each integer between -10 and 10.

Press	Result
[MODE] [▼] [▼] [▼] [ENTER] (sets function graphing mode)	**Normal** Sci Eng **Float** 0123456789 **Radian** Degree **Func** Par Pol Seq **Connected** Dot **Sequential** Simul **Real** a+bi re^θi **Full** Horiz G-T
[Y=]	Plot1 Plot2 Plot3 \Y₁=■ \Y₂= \Y₃= \Y₄= \Y₅= \Y₆= \Y₇=
[X,T,Θ,n] [MATH] 3 [−] 2 [X,T,Θ,n]	Plot1 Plot2 Plot3 \Y₁=X³-2X■ \Y₂= \Y₃= \Y₄= \Y₅= \Y₆= \Y₇=
[2nd] [TBLSET]	TABLE SETUP TblStart=0 △Tbl=1 Indpnt: **Auto** Ask Depend: **Auto** Ask

28

Building a table (continued)

Press	Result
[(-)] 1 0 [ENTER] (sets TblStart; default settings shown for the other fields are appropriate)	TABLE SETUP TblStart=-10 △Tbl=1 Indpnt: **Auto** Ask Depend: **Auto** Ask
[2nd] [TABLE]	X | Y₁ -10 | -980 -9 | -711 -8 | -496 -7 | -329 -6 | -204 -5 | -115 -4 | -56 X=-10

Note: Press [▼] repeatedly to see the changes in X and Y.

Clearing the Y= Editor

Before proceeding with the remaining examples in this guide, clear the Y= Editor.

Press	Result
[Y=]	Plot1 Plot2 Plot3 \Y₁■X³-2X■ \Y₂= \Y₃= \Y₄= \Y₅= \Y₆= \Y₇=
[CLEAR]	Plot1 Plot2 Plot3 \Y₁=■ \Y₂= \Y₃= \Y₄= \Y₅= \Y₆= \Y₇=

Using the CATALOG

The CATALOG is an alphabetic list of all functions and instructions on the TI-83 Plus. Some of these items are also available on keys and menus.

For more details, see Guidebook Chapter 15.

To select from the CATALOG:

1. Position the cursor where you want to insert the item.
2. Press [2nd] [CATALOG].
3. Press ▼ or ▲ to move the ▶ indicator to the function or instruction. (You can move quickly down the list by typing the first letter of the item you need.)
4. Press [ENTER]. Your selection is pasted on the home screen.

   Notes:
   - *Items are listed in alphabetical order. Those that do not start with a letter (+, ≥, √, π, and so on) are at the end of the list.*
   - *You can also paste from the CATALOG to an editor, such as the Y= Editor.*

Example: Enter the *rand* function on the home screen.

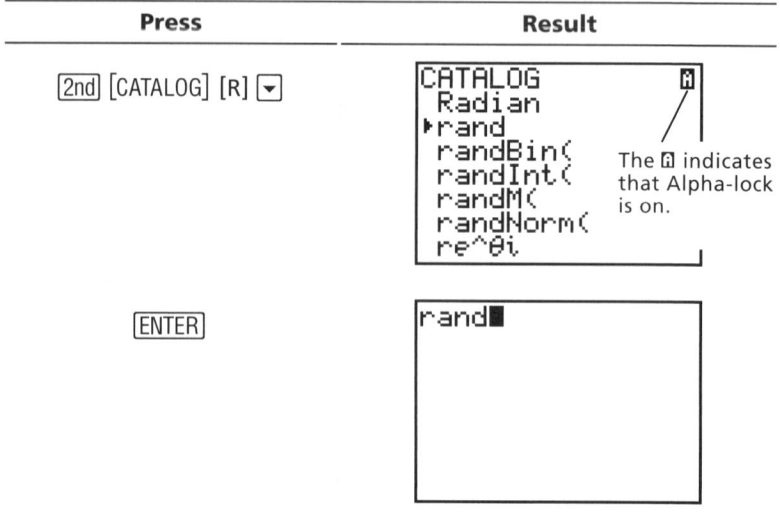

30

Performing simple calculations

Changing a decimal to a fraction

Example: Add 1/2 + 1/4 and change your answer to a fraction.

Press	Result
1 ÷ 2 + 1 ÷ 4 [ENTER]	1/2+1/4 .75
[MATH] 1 [ENTER]	1/2+1/4 .75 Ans▶Frac 3/4

Finding the least common multiple

Example: Find the least common multiple of 15 and 24.

Press	Result
[MATH] ▶ ▲ ▲ [ENTER] 1 5 , 2 4) [ENTER]	lcm(15,24) 120

Performing simple calculations (continued)

Finding the square root

Example: Find the square root of 256.

Press	Result
[2nd] [√] 2 5 6 [)] [ENTER]	√(256) 16

Finding the factorial of numbers

Example: Compute the factorial of 5 and 30.

Press	Result
5 [MATH] [▶] [▶] [▶] 4 [ENTER]	5! 120

3 0 [MATH] [▶] [▶] [▶] 4 [ENTER]	5!
	           120
	30!
	  2.652528598ᴇ32 ← Scientific notation

Performing simple calculations (continued)

Solving trigonometric functions

Example: Find the sine of an angle of 72°.

Press	Result
[SIN] 7 2 [2nd] [ANGLE] [ENTER] [)] [ENTER]	sin(72°) .9510565163

Note: *If you are solving multiple problems using angles, be sure that mode is set to Degree. If you are in Radian mode and do not wish to change the mode, you can use [2nd] [ANGLE] [ENTER] (as you did in this example) to add the degree symbol to the calculation and override the Radian mode setting.*

Adding Complex Numbers

Example: Add (3+5i) + (2-3i).

Press	Result
[(] 3 [+] 5 [2nd] [i] [)] [+] [(] 2 [−] 3 [2nd] [i] [)] [ENTER]	(3+5i)+(2-3i) 5+2i

Note: *The i character is the second function of [.] (the decimal key).*

Using the equation solver

You can use the TI-83 Plus equation solver to solve for a variable in an equation.

Example: Find the roots for the equation $X^2 - 13X - 48 = 0$.

Press	Result

[MATH] [▲]	`MATH NUM CPX PRB`
	`4↑³√(`
	`5: ˣ√`
	`6: fMin(`
	`7: fMax(`
	`8: nDeriv(`
	`9: fnInt(`
	`0: Solver...`

[ENTER]	`EQUATION SOLVER` `eqn:0=`

Note: If you do not see **eqn:0=** as shown above, press [▲] (the up arrow), and then press [CLEAR] to erase the existing equation.

[X,T,Θ,n] [x²] [−]

1 3 [X,T,Θ,n] [−] 4 8

`EQUATION SOLVER`
`eqn:0=X²-13X-48`

34

Using the equation solver (continued)

Press	Result
[ENTER]	X²-13X-48=0 X=0 bound={-1E99,1...
[ALPHA] [SOLVE]	X²-13X-48=0 • X=-3 bound={-1E99,1... • left-rt=0
1 0 0	X²-13X-48=0 X=100■ bound={-1E99,1... left-rt=0
[ALPHA] [SOLVE]	X²-13X-48=0 • X=16 bound={-1E99,1... • left-rt=0

The two roots are -3 and 16. Since you did not enter a guess, the TI-83 Plus used 0 (the default guess) and first returned the answer nearest 0. To find other roots, you must enter another guess. In this example, you entered 100.

35

Entering data into lists

You can enter data into lists using either of two methods:

For more details, see Guidebook Chapter 11 and Chapter 12.

- Using braces and [STO▸] on the home screen

— or —

- Using the statistical list editor.

Using [STO▸]

Example: Store 1, 2, 3, and 4 to list 1 (L1).

Press	Result
[2nd] [{] 1 [,] 2 [,] 3 [,] 4 [2nd] [}]	{1,2,3,4}
[STO▸]	{1,2,3,4}→■
[2nd] [L1] [ENTER]	{1,2,3,4}→L1 {1 2 3 4}

36

Entering data into lists (continued)

Using the statistical list editor

Example: Store 5, 6, 7, and 8 to list 2 (L2).

Press	Result
[STAT] [ENTER]	
[▶] [▲] [CLEAR] [ENTER] (if L2 already contains data)	
5 [ENTER] 6 [ENTER] 7 [ENTER] 8 [ENTER]	
[2nd] [QUIT] [2nd] [L2] [ENTER] (displays the contents of the list on the home screen)	{5 6 7 8}

37

Plotting data

When you have statistical data stored in lists, you can display the data you have collected in a scatter plot, xyLine, histogram, box plot, or normal probability plot.

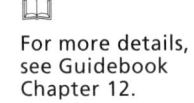

For more details, see Guidebook Chapter 12.

You will need to:

1. Determine which lists contain your data.
2. Tell the calculator which lists of data you want to plot and define the plot.
3. Display the plot.

Determine which lists contain your data

Press	Result
[STAT]	EDIT CALC TESTS 1:Edit... 2:SortA(3:SortD(4:ClrList 5:SetUpEditor
[ENTER]	L1 L2 L3 2 1 5 2 6 3 7 4 8 ------ ------ L2(1)=5

Note: In some cases, you may have several lists stored and you may have to press [▶] *several times to find the correct lists.*

Plotting data (continued)

Tell the calculator which lists you want to plot

Press	Result

[2nd] [STAT PLOT]	**STAT PLOTS**
	1:Plot1...On
	   L1 L2
	2:Plot2...On
	   L1 L2
	3:Plot3...Off
	   L1 L2
	4↓PlotsOff

4 [ENTER] (turns plots off if any plots are on)	PlotsOff Done

[2nd] [STAT PLOT]	**STAT PLOTS**
	1:Plot1...Off
	   L1 L2
	2:Plot2...Off
	   L1 L2
	3:Plot3...Off
	   L1 L2
	4↓PlotsOff

[ENTER]	**Plot1** Plot2 Plot3 On **Off** Type: ▦ ⌐ ⊪ ⊞ ⊞ ⌐ Xlist:L1 Ylist:L2 Mark: ■ + ·

Plotting data (continued)

Press	Result
[ENTER] (turns Plot1 on)	Plot1 **On** Off Type: ... Xlist:L1 Ylist:L2 Mark: ▫ + ·
▼ ▼ [2nd] [LIST] [ENTER] (enters L1 as the Xlist)	Plot1 **On** Off Type: ... Xlist:L1 Ylist:L2 Mark: ▫ + ·
▼ [2nd] [LIST] ▼ [ENTER] (enters L2 as the Ylist)	Plot1 **On** Off Type: ... Xlist:L1 Ylist:L2 Mark: ▫ + ·
▼ ▶ [ENTER] (selects + as the plotting mark)	Plot1 **On** Off Type: ... Xlist:L1 Ylist:L2 Mark: ▫ + ·

Plotting data (continued)

Press	Result
[Y=] [CLEAR]	**Plot1** Plot2 Plot3 \Y₁=■ \Y₂= \Y₃= \Y₄= \Y₅= \Y₆= \Y₇=

Note: *This step is optional and is not necessary unless there is a previous entry in the Y= Editor. If there are additional entries in the Y= Editor, press [▼] [CLEAR] until all are clear.*

Display the plot

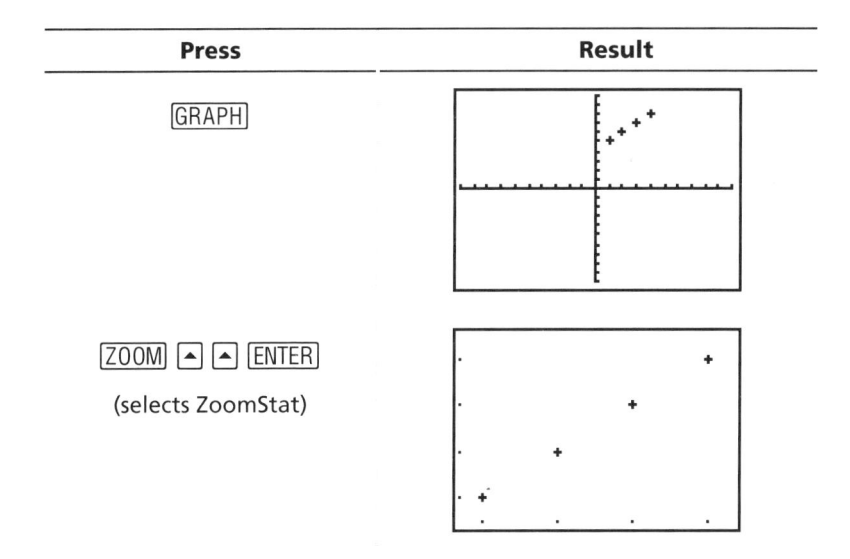

Press	Result
[GRAPH]	(scatter plot)
[ZOOM] [▲] [▲] [ENTER] (selects ZoomStat)	(scatter plot)

Note: *If you would like to add the regression line to a scatter plot, follow the instructions on page 42, adding Y1 to the end of the instruction:*
LinReg(ax+b) L1, L2, Y1. *(Press [VARS] [▶] [ENTER] [ENTER] to add Y1.) Press [GRAPH] to see the regression line.*

Calculating a linear regression

If you wish to calculate the linear regression for data, you can do so using the **LinReg** instruction from the [STAT] CALC menu.

Example: Calculate the linear regression for the data entered in L1 and L2 (on pages 36 and 37).

Press	Result
[STAT] [▶] [▼] [▼] [▼]	EDIT **CALC** TESTS 1: 1-Var Stats 2: 2-Var Stats 3: Med-Med 4: LinReg(ax+b) 5: QuadReg 6: CubicReg 7↓QuartReg
[ENTER]	LinReg(ax+b) ■
[2nd] [L1] [,] [2nd] [L2]	LinReg(ax+b) L1, L2■

[ENTER]	LinReg
	  y=ax+b
	  a=1
	  b=4

Note: The information on the last screen means that the points in L1 and L2 [(1,5) (2,6) (3,7) (4,8)] all lie on the line Y = X + 4.

Calculating statistical variables

The TI-83 Plus lets you easily calculate one-variable or two-variable statistics for data that you have entered into lists.

Example: Using the data that you entered into L1 on page 36, calculate one-variable statistics.

Press	Result
[STAT] [▶]	EDIT **CALC** TESTS **1:**1-Var Stats 2:2-Var Stats 3:Med-Med 4:LinReg(ax+b) 5:QuadReg 6:CubicReg 7↓QuartReg
[ENTER]	1-Var Stats
[2nd] [L1]	1-Var Stats L1
[ENTER]	1-Var Stats x̄=2.5 Σx=10 Σx²=30 Sx=1.290994449 σx=1.118033989 ↓n=4

43

Using the MATRIX Editor

Creating a new matrix

For more details, see Guidebook Chapter 10.

Press	Result
[2nd] [MATRIX] ◄	NAMES MATH **EDIT** 1: [A] 2: [B] 3: [C] 4: [D] 5: [E] 6: [F] 7↓[G]
[ENTER]	MATRIX[A] 1 ×1 [0]
2 [ENTER] 2 [ENTER]	MATRIX[A] 2 ×2 [0 0] [0 0] 1,1=0
1 [ENTER] 5 [ENTER] 2 [ENTER] 8 [ENTER]	MATRIX[A] 2 ×2 [1 5] [2 8] 2,2=8

Note: When you press [ENTER], the cursor automatically highlights the next cell so that you can continue entering or editing values. To enter a new value, you can start typing without pressing [ENTER], but you must press [ENTER] to edit an existing value.

44

Using the MATRIX Editor (continued)

Using matrices to solve systems of equations

You can solve several equations simultaneously by entering their coefficients into a matrix and then using the **rref** (reduced row-echelon form) function. For example, in the equations below, enter 3, 3, and 24 (for 3X, 3Y, and 24) in the first row, and 2, 1, 13 (for 2X, 1Y, and 13) in the second row.

Example: Solve $3X + 3Y = 24$
and $2X + Y = 13$

Press	Result
[2nd] [MATRIX] ▶ ▶ ▼	NAMES MATH **EDIT** 1:[A] 2×2 2:[B] 3:[C] 4:[D] 5:[E] 6:[F] 7↓[G]
[ENTER]	MATRIX[B] 1 ×1 [0]
2 [ENTER] 3 [ENTER]	MATRIX[B] 2 ×3 [0 0 0] [0 0 0] 1,1=0

45

Using the MATRIX Editor (continued)

Press	Result
3 [ENTER] 3 [ENTER] 2 4 [ENTER]	MATRIX[B] 2 ×3 [3 3 24] [2 1 ▓]
2 [ENTER] 1 [ENTER] 1 3 [ENTER]	2,3=13
[2nd] [QUIT]	∎
[2nd] [MATRIX] [▶]	NAMES **MATH** EDIT 1:det(2:T 3:dim(4:Fill(5:identity(6:randM(7↓augment(
[▲] [▲] [▲] [▲] [▲]	NAMES **MATH** EDIT 0↑cumSum(A:ref(B:rref(C:rowSwap(D:row+(E:*row(F:*row+(
[ENTER]	rref(

46

Using the MATRIX Editor (continued)

Press	Result
[2nd] [MATRIX] [▼] [ENTER]	rref([B]■
[ENTER]	rref([B] [[1 0 5] [0 1 3]]

You can interpret the resulting matrix as:

[1 0 5] represents 1X + 0Y = 5 or X = 5

[0 1 3] represents 0X + 1Y = 3 or Y = 3

The solution to this system of equations is X = 5, Y = 3.

Grouping

Grouping lets you make a copy of two or more variables and store them in the Flash memory of the TI-83 Plus. This function is similar to "zipping" a computer file and storing it. For example, suppose that you want to save data you collected for time, temperature, humidity, and barometric pressure because you may need to use the data for another assignment.

For more details, see Guidebook Chapter 18.

Grouping lets you keep these lists together for future use. Instead of trying to locate the correct lists and remember which ones were collected together, you can simply recall the group. Grouping also saves space on your calculator by copying variables from RAM to Flash memory.

Example: Group lists L1, L2, and L3 and name them GROUPA.

Press	Result
[2nd] [MEM]	**MEMORY** 2:Mem Mgmt/Del… 3:Clear Entries 4:ClrAllLists 5:Archive 6:UnArchive 7:Reset… 8:Group…
8	**GROUP** UNGROUP 1:Create New
[ENTER]	GROUP Name=▮ *You are in alpha mode.*

48

Grouping (continued)

Press	Result
[G] [R] [O] [U] [P] [A]	GROUP Name=GROUPA
[ENTER]	GROUP 1:All+... 2:All-... 3:Prgm... 4:List... 5:GDB... 6:Pic... 7↓Matrix...

4	SELECT DONE
	▶ L₁ LIST
	   L₂ LIST
	   L₃ LIST
	   L₄ LIST
	   L₅ LIST
	   L₆ LIST

[ENTER] [▼] [ENTER]
[▼] [ENTER]
	SELECT DONE
	■ L₁ LIST
	■ L₂ LIST
	■ L₃ LIST
	♦ L₄ LIST
	   L₅ LIST
	   L₆ LIST

[▶]	SELECT DONE 1:Done
[ENTER]	Copying Variables to Group: GROUPA Done

Ungrouping

To use variables that have been grouped, you must ungroup.

Example: Ungroup GROUPA (which you grouped on page 50).

Press	Result
[2nd] [MEM]	**MEMORY** 2↑Mem Mgmt/Del… 3:Clear Entries 4:ClrAllLists 5:Archive 6:UnArchive 7:Reset… 8:Group…
8	**GROUP** UNGROUP 1:Create New
[▶]	GROUP **UNGROUP** 1:*GROUPA
[ENTER]	**DuplicateName** 1:Rename 2:Overwrite 3:Overwrite All 4:Omit 5:Quit L₁ LIST

3
(to overwrite all three lists)
	Ungrouping:
	GROUPA
	  L₁ LIST
	  L₂ LIST
	▶ L₃ LIST
	              Done

Error messages

Occasionally, when you enter a function or instruction or attempt to display a graph, the TI-83 Plus will return an error message.

For more details, see Guidebook Appendix B.

Example: Enter the least common multiple function **lcm(** followed by only one number.

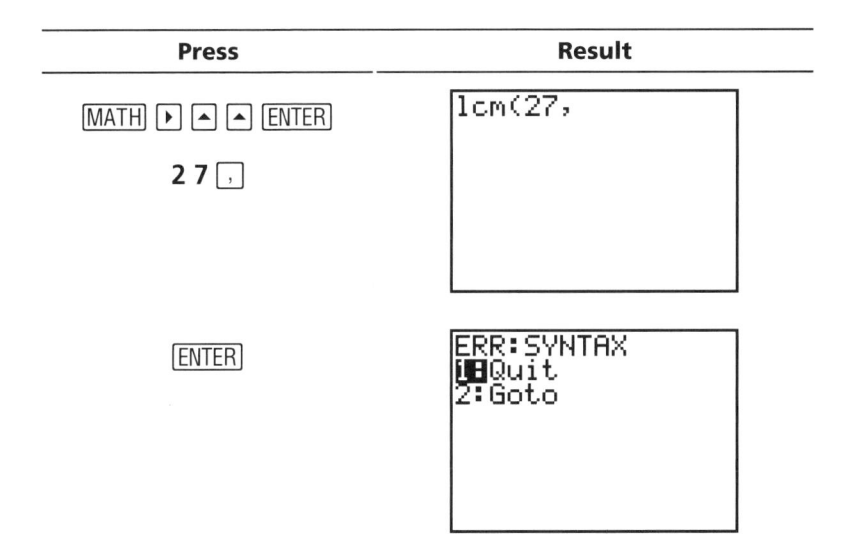

Press	Result
[MATH] [▶] [▲] [▲] [ENTER] 2 7 [,]	`lcm(27,`

[ENTER]	`ERR:SYNTAX`
	`1:Quit`
	`2:Goto`

If you select **1:Quit**, you return to the home screen with the cursor on a new entry line. If you select **2:Goto**, you return to the original entry line; the cursor is flashing at the location of the error. You can now correct the error and continue.

You can find a complete list of error conditions with explanations in the Guidebook, Appendix B: General Information.

Resetting defaults

If you are getting unexpected results, or if another person has used your calculator and may have changed the settings, you should consider resetting defaults on the TI-83 Plus.

For more details, see Guidebook Chapter 18.

Press	Result
[2nd] [MEM]	**MEMORY** 1: About 2: Mem Mgmt/Del… 3: Clear Entries 4: ClrAllLists 5: Archive 6: UnArchive 7↓Reset…
7	RAM **ARCHIVE** ALL 1: All RAM… 2: Defaults…
2	**RESET DEFAULTS** 1: No 2: Reset
2	■ TI-83 Plus 1.19 Defaults set

WARNING: *If you reset All RAM in step 3 above, you will delete stored variables, lists, applications, and programs. Be sure you have backed up any essential data before you select this option.*

Installing applications

Calculator software applications (APPS) let you update the functionality of your TI-83 Plus by installing APPS. This is similar to the way that you add new features to your computer by installing new software applications.

You can find applications for the TI-83 Plus at the TI Online Store at **education.ti.com**. Once you have downloaded an application to your computer, you must use TI Connect™ or TI-GRAPH LINK™ software and TI-GRAPH LINK cable to install the application on your TI-83 Plus calculator.

Instructions for Windows®

1. Connect the TI-GRAPH LINK cable between your computer and calculator. Make sure the calculator is on the home screen.
2. Using Windows (or NT) Explorer, locate the application file you want to transfer to the connected device.
3. Reduce the size of the Explorer window so you can see the TI Connect desktop icon.
4. Click the application file you want to transfer.
5. Drag the application file out of Explorer and drop it on the TI Connect desktop icon.

Instructions for Macintosh®

1. Connect the TI-GRAPH LINK cable between your computer and calculator, and make sure the calculator is on the home screen.
2. Launch the TI-GRAPH LINK 2 software and establish a connection to your calculator.
3. Drag the application to the calculator window in TI-GRAPH LINK. Follow any on-screen instructions that are given.

Running applications

Once you have installed an application on your TI-83 Plus, you must start the application to use its features.

Example: Start the Catalog Help (CtlgHelp) app on the TI-83 Plus.

Press	Result
[APPS]	APPLICATIONS 1:Finance… 2:CBL/CBR 3:CtlgHelp 4:Organize 5:Periodic 6:Prob Sim 7↓PuzzPack
▼ ▼ [ENTER]	**TI-83Plus** Catalog Help 1.0 ©Texas Instruments 2000

Quick reference

Press	To
[2nd] [▲]	Darken the screen
[2nd] [▼]	Lighten the screen
[2nd] [▶]	Move the cursor to the end of an expression
[2nd] [◀]	Move the cursor to the beginning of an expression
[ALPHA] [▼]	Page down to the next screen (on menus)
[ALPHA] [▲]	Page up to the next screen (on menus)
[2nd] [ENTRY]	Place your last entry on the current entry line on the home screen
[2nd] [ANS]	Place Ans (a reference to your last answer) on the current entry line on the home screen, allowing you to use the answer in the next calculation
[DEL]	Delete the character under the cursor
[2nd] [INS]	Insert additional characters at the cursor
[▼] [▲]	Move the cursor from line to line
[▶] [◀]	Move the cursor from character to character within a line
[CLEAR]	Clear the current line. (If the cursor is on a blank line, clears everything on the home screen.)

Texas Instruments (TI) Support and Service

For General Information

Home Page:	education.ti.com
KnowledgeBase and E-mail Inquiries:	education.ti.com/support
Phone:	(800) TI-CARES; (800) 842-2737 For U.S., Canada, Mexico, Puerto Rico, and Virgin Islands only
International Information:	education.ti.com/international

For Technical Support

KnowledgeBase and Support by E-mail:	education.ti.com/support
Phone (not toll-free):	(972) 917-8324

For Product (hardware) Service

Customers in the U.S., Canada, Mexico, Puerto Rico and Virgin Islands: Always contact TI Customer Support before returning a product for service.

All other customers: Refer to the leaflet enclosed with this product (hardware) or contact your local TI retailer/distributor.